Ten Patterns

AF693700

Example

Find the sum.

40 + 10

Think: **4** + **1** = **5**

So, **40** + **10** = **50**.

Look for a pattern when you add tens.

Find the sum.

1. 10 + 10
2. 20 + 60
3. 50 + 20
4. 40 + 70
5. 30 + 20
6. 50 + 40
7. 80 + 50
8. 40 + 20
9. 60 + 60
10. 20 + 20
11. Emily sold 90 tickets in one week. The second week she sold 10 more tickets. How many tickets did she sell altogether?
12. There are 20 people on the roller coaster. Then 10 more get on. How many are on the roller coaster?

Answer Box

A	B	C	D	E	F
90	60	110	40	70	100
G	**H**	**I**	**J**	**K**	**L**
30	80	130	20	50	120

Objective: Find the sum of tens.

Hundred Patterns

Example

Find the sum.

400 + 100

Think: **4** + **1** = **5**

So, **4**00 + **1**00 = **5**00.

Find the sum.

1. 200 + 300
2. 200 + 100
3. 100 + 100
4. 700 + 500
5. 400 + 600
6. 100 + 300
7. 500 + 200
8. 900 + 200
9. 600 + 700
10. 600 + 300
11. Lake School has 500 students. Hillside School has 300 students. How many students are there altogether?
12. At Central School, 400 students ride the bus. The other 200 students walk. How many students are at Central School?

Answer Box

A	B	C	D	E	F
600	1,100	1,000	800	700	900
G	**H**	**I**	**J**	**K**	**L**
300	400	1,300	200	500	1,200

 Objective: Find the sum of hundreds.

It's Great to Estimate!

To estimate a sum, first round the addends.

To round to the nearest ten, look at the digit in the ones place.

To round to the nearest hundred, look at the digit in the tens place.

24 + 67	Round each addend to the nearest ten.	20 + 70 = 90
352 + 428	Round each addend to the nearest hundred.	400 + 400 = 800

Find the best estimate.

1. 46 + 12
2. 621 + 198
3. 26 + 43
4. 42 + 11
5. 689 + 235
6. 56 + 32
7. 54 + 29
8. 485 + 186
9. 236 + 195
10. 114 + 526
11. 269 + 240
12. 58 + 41

Answer Box

A	B	C	D	E	F
900	500	90	800	80	600
G	**H**	**I**	**J**	**K**	**L**
70	400	100	700	50	60

Objective: Find the estimated sum of two whole numbers.

Let's Add Two Numbers!

Find the sum.

1	2	3	4	5
35 + 21	14 + 61	31 + 48	76 + 23	13 + 25

6	7	8	9	10
23 + 30	40 + 57	45 + 12	62 + 26	55 + 43

11. Brent has 36 photos of mountains. He has 41 photos of lakes and rivers. How many photos does he have in all?

12. Nina can take 24 pictures with one roll of film. She can take 12 pictures with another roll. How many pictures can she take altogether?

Answer Box

A	B	C	D	E	F
79	88	99	56	53	75
G	**H**	**I**	**J**	**K**	**L**
38	77	97	98	36	57

 Objective: Find the sum of two 2-digit numbers with no regrouping.

Let's Regroup!

Example

Find the sum.

```
  1
  45    Add the ones. 5 + 7 = 12
+ 27    Regroup 12 ones as 1 ten and 2 ones.
  72    Add the tens. 1 + 4 + 2 = 7
```

Find the sum.

1. 43 + 48
2. 26 + 34
3. 19 + 67
4. 44 + 26
5. 46 + 35
6. 18 + 37
7. 39 + 25
8. 26 + 19
9. 35 + 16
10. 19 + 35
11. Amber has 24 lilies in her garden. She has 17 roses. How many flowers does Amber have in all?
12. A flower shop has 28 plants. A shipment of 68 more plants arrives. Now how many plants does the shop have?

Answer Box

A	B	C	D	E	F
91	70	60	55	51	86
G	**H**	**I**	**J**	**K**	**L**
54	64	45	96	41	81

Objective: Find the sum of two 2-digit numbers with regrouping.

You Decide!

Find the sum.

1. 27 + 49	2. 41 + 27	3. 17 + 25	4. 19 + 37	5. 39 + 21	6. 68 + 16
7. 22 + 28	8. 47 + 26	9. 16 + 42	10. 17 + 57	11. 48 + 18	12. 26 + 6

Answer Box

A	B	C	D	E	F
50	74	60	66	32	73
G	**H**	**I**	**J**	**K**	**L**
56	76	42	58	84	68

Find the sum.

1. 42 + 36	2. 14 + 58	3. 69 + 27	4. 35 + 35	5. 19 + 76	6. 18 + 42
7. 43 + 19	8. 39 + 46	9. 55 + 13	10. 16 + 18	11. 11 + 13	12. 68 + 18

Answer Box

A	B	C	D	E	F
60	72	34	62	96	68
G	**H**	**I**	**J**	**K**	**L**
86	85	78	70	95	24

Objective: Find the sum of two 2-digit numbers with regrouping.

Hundreds and More Hundreds

Example

Find the sum.

$$\begin{array}{r} {}^{1} \\ 361 \\ \underline{+\ 172} \\ \mathbf{533} \end{array}$$

Add the ones. 1 + 2 = 3
Add the tens. 6 + 7 = 13
Regroup 13 tens as 1 hundred and 3 tens.
Add the hundreds. 1 + 3 + 1 = 5

Find the sum.

1. $\begin{array}{r} 322 \\ \underline{+\ 596} \end{array}$

2. $\begin{array}{r} 136 \\ \underline{+\ 237} \end{array}$

3. $\begin{array}{r} 289 \\ \underline{+\ 502} \end{array}$

4. $\begin{array}{r} 478 \\ \underline{+\ \ \ 71} \end{array}$

5. $\begin{array}{r} 335 \\ \underline{+\ 647} \end{array}$

6. $\begin{array}{r} 541 \\ \underline{+\ 395} \end{array}$

7. $\begin{array}{r} 368 \\ \underline{+\ 105} \end{array}$

8. $\begin{array}{r} 439 \\ \underline{+\ 533} \end{array}$

9. $\begin{array}{r} 290 \\ \underline{+\ 491} \end{array}$

10. $\begin{array}{r} 175 \\ \underline{+\ 284} \end{array}$

11. $\begin{array}{r} 327 \\ \underline{+\ 509} \end{array}$

12. $\begin{array}{r} 886 \\ \underline{+\ \ \ 42} \end{array}$

Answer Box

A	B	C	D	E	F
928	972	791	781	473	459
G	**H**	**I**	**J**	**K**	**L**
373	936	836	982	918	549

Objective: Find the sum of two 3-digit numbers with regrouping.

Moving On with Addition!

Example

Find the sum.

$$\begin{array}{r} {\scriptstyle 1\,1} \\ 326 \\ +\ 198 \\ \hline \mathbf{524} \end{array}$$

Add the ones. 6 + 8 = 14
Regroup 14 ones as 1 ten and 4 ones.
Add the tens. 1 + 2 + 9 = 12
Regroup 12 tens as 1 hundred and 2 tens.
Add the hundreds. 1 + 3 + 1 = 5

Find the sum.

1. $\begin{array}{r} 159 \\ +\ 329 \\ \hline \end{array}$
2. $\begin{array}{r} 275 \\ +\ 162 \\ \hline \end{array}$
3. $\begin{array}{r} 468 \\ +\ 373 \\ \hline \end{array}$
4. $\begin{array}{r} 679 \\ +\ 346 \\ \hline \end{array}$
5. $\begin{array}{r} 361 \\ +\ 190 \\ \hline \end{array}$
6. $\begin{array}{r} 108 \\ +\ 47 \\ \hline \end{array}$
7. $\begin{array}{r} 252 \\ +\ 479 \\ \hline \end{array}$
8. $\begin{array}{r} 189 \\ +\ 289 \\ \hline \end{array}$
9. $\begin{array}{r} 847 \\ +\ 78 \\ \hline \end{array}$
10. $\begin{array}{r} 258 \\ +\ 189 \\ \hline \end{array}$

11. The school store has 205 notebooks with plain covers. It has 246 notebooks with a cartoon on the cover. How many notebooks does the store have altogether?

12. Kevin bought a jumbo notebook with 175 pages. He also bought a smaller notebook with 80 pages. How many pages are there in all?

Answer Box

A	B	C	D	E	F
451	551	478	255	447	731
G	**H**	**I**	**J**	**K**	**L**
841	925	155	437	1,025	488

 Objective: Find the sum of two 3-digit numbers with regrouping.

Lots of Addends

Example

Find the sum.

```
    1
  107
  531
+ 143
  781
```

Add three or more addends the same way you add two addends. First add ones. Then add tens. Then add hundreds.

Find the sum.

1. 11 + 24 + 37
2. 18 + 40 + 34
3. 206 + 38 + 150
4. 637 + 53 + 35
5. 142 + 526 + 200 + 123
6. 234 + 167 + 250 + 12
7. 127 + 243 + 101 + 365
8. 202 + 520 + 35 + 334
9. 15 + 16 + 51
10. 223 + 301 + 137
11. 195 + 63 + 26
12. 242 + 50 + 673 + 24

Answer Box

A 284	B 394	C 836	D 661	E 1,091	F 663
G 82	H 991	I 72	J 725	K 92	L 989

Objective: Find the sum of three or four 2- to 3-digit numbers with regrouping.

Greater Addends

Find the sum.

1. 2,315 + 3,492 + 1,782

2. 1,356 + 2,409 + 632

3. 1,283 + 52 + 426

4. 3,240 + 5,123 + 28 + 231

5. 64 + 33 + 20 + 79

6. 192 + 325 + 344

7. 54 + 27 + 105

8. 2,232 + 701 + 1,564

9. 2,435 + 70 + 3,754

10. 654 + 49 + 6,785

11. 3,498 people enter the theme park during the morning. 4,124 people enter during the afternoon. How many people enter during the morning and afternoon?

12. 630 people rode the water slide in the morning. 4,273 people rode it in the afternoon. 2,346 people rode it in the evening. How many people rode the water slide that day?

Answer Box

A	B	C	D	E	F
8,622	7,488	4,497	7,249	196	6,259
G	**H**	**I**	**J**	**K**	**L**
186	7,589	861	4,397	1,761	7,622

Objective: Find the sum of two, three, or four 3- to 4-digit numbers with regrouping.

More Ten Patterns

Example

Find the difference.

50 – 10

Think: **5** – **1** = **4**

So, **50** – **10** = **40**.

Find the difference.

1. 50 – 30
2. 80 – 20
3. 20 – 10
4. 100 – 10
5. 70 – 70
6. 60 – 20
7. 90 – 20
8. 110 – 10
9. 70 – 40
10. 120 – 10
11. 90 – 10
12. 80 – 30

Answer Box

A	B	C	D	E	F
110	90	0	80	60	50
G	**H**	**I**	**J**	**K**	**L**
20	70	10	30	100	40

Objective: Find the difference of tens.

More Hundred Patterns

Example

Find the difference.

500 – 300

Think: **5 – 3 = 2**

So, **5**00 – **3**00 = **2**00.

Find the difference.

1. 500 – 100
2. 900 – 200
3. 600 – 500
4. 800 – 800
5. 900 – 600
6. 700 – 200
7. 1,100 – 100
8. 1,200 – 100
9. 600 – 400
10. 900 – 100
11. 1,100 – 200
12. 800 – 200

Answer Box

A	B	C	D	E	F
1,100	600	400	0	300	900
G	**H**	**I**	**J**	**K**	**L**
500	700	1,000	800	200	100

 Objective: Find the difference of hundreds.

Close Differences

To estimate a difference, first round the numbers.
To round to the nearest ten, look at the digit in the ones place.
To round to the nearest hundred, look at the digit in the tens place.

63 – 47	Round each number to the nearest ten.	60 – 50 = 10
483 – 157	Round each number to the nearest hundred.	500 – 200 = 300

Find the best estimate.

1. 423 – 298
2. 63 – 32
3. 489 – 101
4. 98 – 21
5. 775 – 225
6. 87 – 53
7. 78 – 19
8. 910 – 398
9. 880 – 745
10. 69 – 18
11. 94 – 15
12. 635 – 320

Answer Box

A 300	B 30	C 600	D 200	E 60	F 100
G 40	H 500	I 400	J 70	K 80	L 50

Objective: Find the estimated difference of two whole numbers.

There's a Difference!

Find the difference.

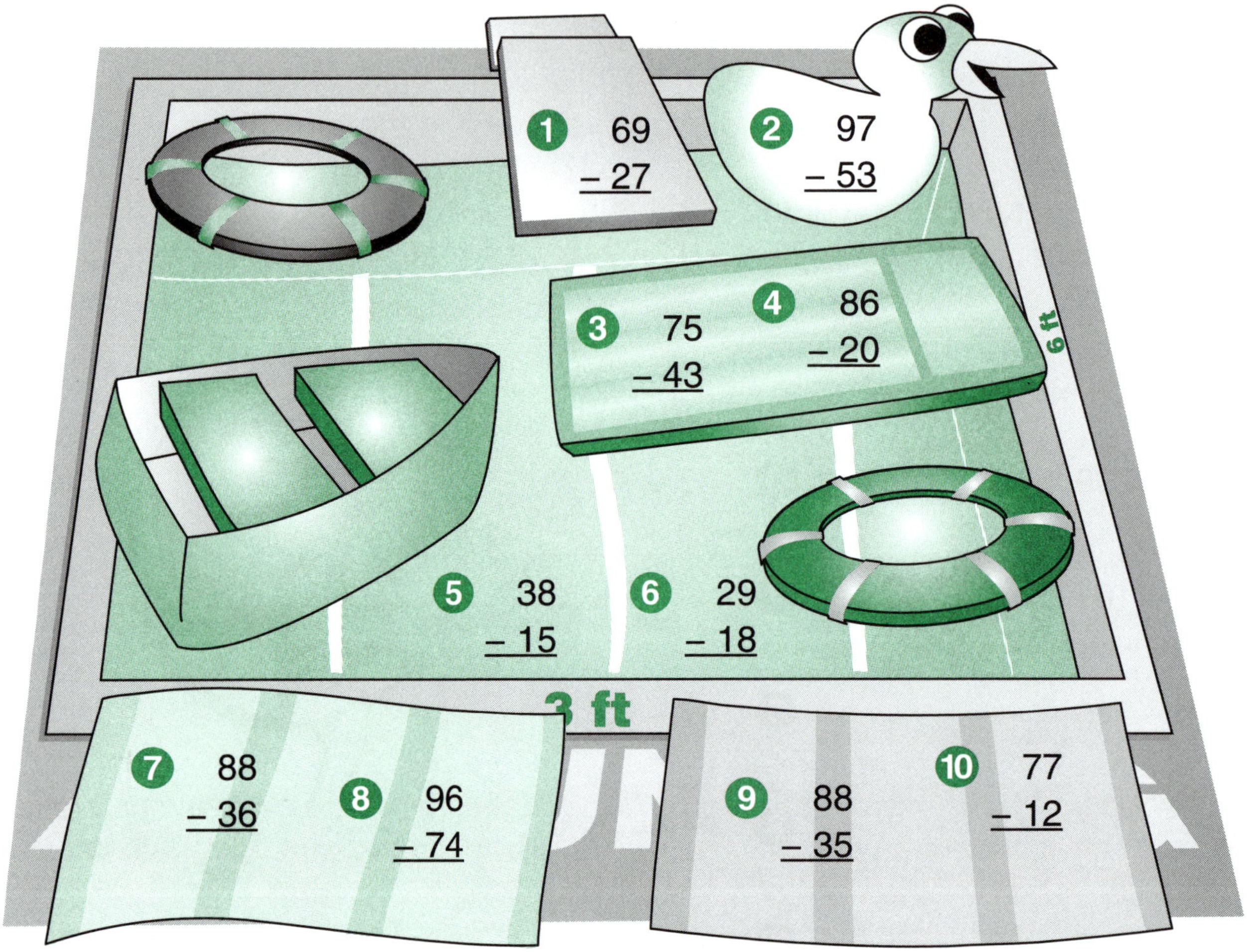

11. 66 children sit at a pool side. 23 children dive in. How many children are not in the pool?

12. 23 children are swimming in the pool. 11 children get out for a snack. How many children are still in the pool?

Answer Box

A	B	C	D	E	F
32	53	22	12	42	66
G	**H**	**I**	**J**	**K**	**L**
43	23	11	44	52	65

 Objective: Find the difference of two 2-digit numbers with no regrouping.

Regroup and Subtract!

Example

Find the difference.

5 12 ~~6~~ ~~2~~ – 3 8 **2 4**	Regroup 6 tens and 2 ones as 5 tens and 12 ones. Subtract the ones. 12 – 8 = 4 Subtract the tens. 5 – 3 = 2

Find the difference.

1. 70 – 19
2. 76 – 37
3. 44 – 29
4. 95 – 48
5. 48 – 19
6. 73 – 36
7. 26 – 18
8. 51 – 16
9. 60 – 42
10. 97 – 38
11. Luis had 52 model cars in his collection. He gave away 27 cars. How many does he have left?
12. Cora had 58 train cars in her collection. She gave away 9 cars. How many does she have left?

Answer Box

A	B	C	D	E	F
25	29	51	8	35	49
G	**H**	**I**	**J**	**K**	**L**
15	18	59	47	37	39

Objective: Find the difference of two 2-digit numbers with regrouping.

Look at the Difference!

Find the difference.

1. 43 − 29	2. 72 − 31	3. 62 − 37	4. 74 − 48	5. 80 − 57	6. 81 − 46
7. 57 − 48	8. 90 − 59	9. 36 − 17	10. 52 − 28	11. 99 − 65	12. 46 − 9

Answer Box

A	B	C	D	E	F
14	26	35	41	19	25
G	**H**	**I**	**J**	**K**	**L**
24	9	37	31	34	23

Find the difference.

1. 42 − 16	2. 55 − 18	3. 63 − 17	4. 75 − 52	5. 50 − 33	6. 48 − 19
7. 83 − 50	8. 64 − 37	9. 75 − 39	10. 73 − 19	11. 81 − 42	12. 68 − 21

Answer Box

A	B	C	D	E	F
54	33	39	17	27	47
G	**H**	**I**	**J**	**K**	**L**
26	23	36	46	37	29

Objective: Find the difference of two 2-digit numbers with regrouping.

Driving Differences

Example

Find the difference.

```
 3 13
 4 3 8
-2 6 6
 1 7 2
```

(4 and 3 crossed out; regrouped as 3 and 13)

Subtract the ones. 8 – 6 = 2
Regroup 4 hundreds and 3 tens as 3 hundreds and 13 tens.
Subtract the tens. 13 – 6 = 7
Subtract the hundreds. 3 – 2 = 1

Find the difference.

1. 847 – 274
2. 784 – 490
3. 663 – 125
4. 976 – 649
5. 739 – 191
6. 522 – 208
7. 764 – 419
8. 816 – 483
9. 915 – 493
10. 892 – 219

11. Leon is driving 526 miles to Chicago. He drove 144 miles in the morning. How many miles does he still have to drive?

12. Annie is driving 463 miles to New York City. She has 128 miles left to drive. How many miles has she already driven?

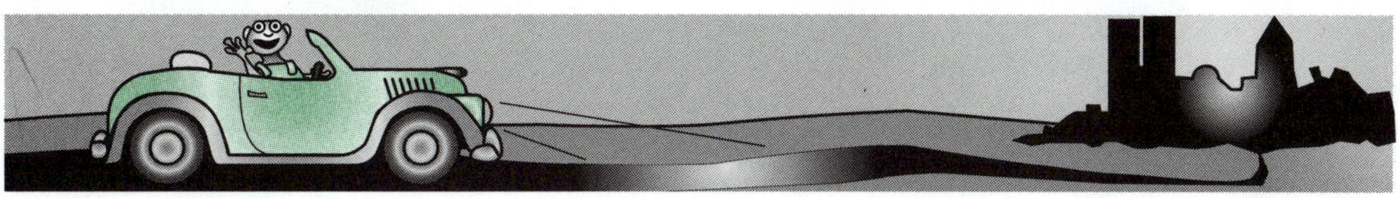

Answer Box

A	B	C	D	E	F
294	314	327	573	422	538
G	**H**	**I**	**J**	**K**	**L**
335	333	673	345	548	382

Objective: Find the difference of two 3-digit numbers with regrouping once.

Make a Difference!

Find the difference.

1. $363 - 127$
2. $592 - 276$
3. $748 - 456$
4. $891 - 583$
5. $439 - 367$
6. $714 - 592$
7. $951 - 629$
8. $667 - 385$
9. $495 - 249$
10. $832 - 514$
11. $655 - 483$
12. $786 - 394$

Answer Box

A	B	C	D	E	F
122	316	172	292	308	236

G	H	I	J	K	L
282	392	246	72	318	322

Find the difference.

1. $718 - 292$
2. $945 - 406$
3. $858 - 487$
4. $281 - 118$
5. $972 - 327$
6. $649 - 371$
7. $783 - 128$
8. $824 - 393$
9. $237 - 64$
10. $642 - 116$
11. $518 - 280$
12. $835 - 294$

Answer Box

A	B	C	D	E	F
173	645	541	431	426	655

G	H	I	J	K	L
238	371	539	278	526	163

Objective: Find the difference of two 3-digit numbers with regrouping once.

Regroup and Regroup Again!

Example

Find the difference.

```
  1 16
 4 2 6   Regroup 2 tens and 6 ones
-1 7 8     as 1 ten and 16 ones.
     8   Subtract the ones.
           16 – 8 = 8
```

```
 3 11
   1 16
 4 2 6   Regroup 4 hundreds and
-1 7 8     1 ten as 3 hundreds and
 2 4 8     11 tens.
         Subtract the tens. 11 – 7 = 4
         Subtract the hundreds. 3 – 1 = 2
```

Find the difference.

1. 537 − 268	2. 626 − 348	3. 725 − 486	4. 846 − 359	5. 638 − 249	6. 924 − 387
7. 772 − 373	8. 842 − 195	9. 634 − 165	10. 943 − 346	11. 955 − 567	12. 833 − 464

Answer Box

A	B	C	D	E	F
389	469	597	399	537	647
G	**H**	**I**	**J**	**K**	**L**
239	388	487	269	369	278

Objective: Find the difference of two 3-digit numbers with regrouping twice.

Once or Twice

Find the difference.

1. $\begin{array}{r} 342 \\ -\ 136 \\ \hline \end{array}$

2. $\begin{array}{r} 515 \\ -\ 298 \\ \hline \end{array}$

3. $\begin{array}{r} 578 \\ -\ 492 \\ \hline \end{array}$

4. $\begin{array}{r} 431 \\ -\ 285 \\ \hline \end{array}$

5. $\begin{array}{r} 712 \\ -\ 447 \\ \hline \end{array}$

6. $\begin{array}{r} 368 \\ -\ 139 \\ \hline \end{array}$

7. $\begin{array}{r} 683 \\ -\ 497 \\ \hline \end{array}$

8. $\begin{array}{r} 891 \\ -\ 616 \\ \hline \end{array}$

9. $\begin{array}{r} 512 \\ -\ 129 \\ \hline \end{array}$

10. $\begin{array}{r} 929 \\ -\ 683 \\ \hline \end{array}$

11. Anita made 248 belts for a craft show. She sold 119 belts. How many belts does she have left?

12. James made 145 bracelets for a jewelry fair. He has 39 bracelets left. How many bracelets did he sell?

Answer Box

A	B	C	D	E	F
206	186	217	229	129	86
G	**H**	**I**	**J**	**K**	**L**
246	146	275	106	383	265

Objective: Find the difference of two 3-digit numbers with regrouping.

Zeros Hold Their Place

Example

Find the difference.

```
 3 10
 4̸ 0̸ 2
−2 8 1
 1 2 1
```

Subtract the ones. $2 - 1 = 1$
Regroup 4 hundreds and 0 tens as 3 hundreds and 10 tens.
Subtract the tens. $10 - 8 = 2$
Subtract the hundreds. $3 - 2 = 1$

Find the difference.

1. $320 - 118$
2. $806 - 485$
3. $709 - 648$
4. $460 - 147$
5. $808 - 293$
6. $590 - 267$
7. $703 - 542$
8. $840 - 628$
9. $906 - 425$
10. $708 - 375$
11. $690 - 363$
12. $840 - 225$

Answer Box

A	B	C	D	E	F
161	333	481	61	212	615
G	**H**	**I**	**J**	**K**	**L**
202	313	327	515	323	321

Objective: Find the difference of two 3-digit numbers with regrouping across zeros.

Zero Power!

Example

Find the difference.

```
 2 10
 3̸ 0̸ 4
−1  9 6
```

You can't regroup 0 tens and 4 ones, so regroup 3 hundreds and 0 tens as 2 hundreds and 10 tens.

```
    9
 2 1̸0̸ 14
 3̸  0̸  4̸
−1  9  6
 1  0  8
```

Then regroup 10 tens and 4 ones as 9 tens and 14 ones. Then subtract ones, tens, and hundreds.

Find the difference.

1. $\begin{array}{r} 604 \\ -\ 338 \\ \hline \end{array}$
2. $\begin{array}{r} 860 \\ -\ 502 \\ \hline \end{array}$
3. $\begin{array}{r} 709 \\ -\ 517 \\ \hline \end{array}$
4. $\begin{array}{r} 906 \\ -\ 328 \\ \hline \end{array}$
5. $\begin{array}{r} 480 \\ -\ 164 \\ \hline \end{array}$
6. $\begin{array}{r} 550 \\ -\ 185 \\ \hline \end{array}$
7. $\begin{array}{r} 909 \\ -\ 231 \\ \hline \end{array}$
8. $\begin{array}{r} 705 \\ -\ 429 \\ \hline \end{array}$
9. $\begin{array}{r} 500 \\ -\ 208 \\ \hline \end{array}$
10. $\begin{array}{r} 790 \\ -\ 315 \\ \hline \end{array}$
11. $\begin{array}{r} 800 \\ -\ 476 \\ \hline \end{array}$
12. $\begin{array}{r} 508 \\ -\ 140 \\ \hline \end{array}$

Answer Box

A	B	C	D	E	F
365	358	578	266	192	292
G	**H**	**I**	**J**	**K**	**L**
276	368	475	678	324	316

Objective: Find the difference of two 3-digit numbers with regrouping across zeros.

Thousands from Thousands

Find the difference.

1. 7,249 − 3,627
2. 9,418 − 4,270
3. 5,241 − 3,182
4. 8,834 − 6,737
5. 7,398 − 5,329
6. 9,068 − 6,961
7. 6,953 − 4,593
8. 5,820 − 1,198
9. 5,504 − 2,376
10. 8,739 − 3,491
11. 8,560 − 5,448
12. 4,293 − 2,933

Answer Box

A	B	C	D	E	F
5,148	2,107	2,069	3,112	2,097	3,622
G	**H**	**I**	**J**	**K**	**L**
1,360	4,622	3,128	2,059	5,248	2,360

Objective: Find the difference of two 4-digit numbers with regrouping.

Dollars and Cents

Find the sum.

1. $5.62 + 3.27
2. $3.39 + 4.28
3. $2.88 + 6.25
4. $7.08 + 1.96
5. $4.58 + 5.36
6. $5.34 + 3.35
7. $1.42 + 3.70 + 2.41
8. $6.32 + 3.57
9. $2.47 + 2.54 + 3.78
10. $1.79 + 5.78
11. $4.36 + 1.20 + 2.17
12. $2.38 + 5.65

Answer Box

A	B	C	D	E	F
$7.73	$9.94	$9.04	$8.89	$9.89	$8.03
G	**H**	**I**	**J**	**K**	**L**
$9.13	$8.79	$7.57	$7.53	$8.69	$7.67

Find the sum.

1. $3.29 + 4.29
2. $6.26 + 1.95
3. $4.34 + 4.73
4. $5.41 + 2.64
5. $2.29 + 5.35
6. $3.93 + 5.04
7. $2.51 + 4.54
8. $3.43 + 3.68
9. $4.21 + 3.30 + 1.23
10. $1.32 + 2.64 + 4.62
11. $5.32 + 1.36 + 2.84
12. $2.53 + 3.45 + 3.64

Answer Box

A	B	C	D	E	F
$8.05	$7.58	$8.97	$7.11	$9.07	$8.74
G	**H**	**I**	**J**	**K**	**L**
$7.05	$8.58	$9.62	$8.21	$7.64	$9.52

 Objective: Find the sum of two or three amounts of money.

How Much Is Left?

Find the difference.

1. $5.25 − 3.18	2. $6.47 − 3.14	3. $7.36 − 4.18	4. $8.17 − 6.49	5. $4.01 − 2.60	6. $7.24 − 5.38
7. $7.89 − 5.21	8. $8.26 − 4.94	9. $9.81 − 7.67	10. $6.20 − 2.98	11. $4.02 − 2.71	12. $9.16 − 6.99

Answer Box

A	B	C	D	E	F
$2.07	$1.68	$1.41	$1.31	$3.33	$1.86
G	**H**	**I**	**J**	**K**	**L**
$3.22	$2.68	$3.18	$2.14	$3.32	$2.17

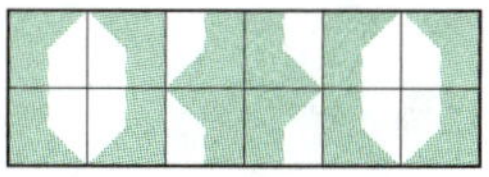

Find the difference.

1. $7.14 − 4.63	2. $9.98 − 3.52	3. $5.83 − 4.27	4. $6.21 − 3.65	5. $4.09 − 1.32	6. $6.15 − 1.47
7. $4.96 − 3.32	8. $7.60 − 4.09	9. $7.92 − 2.14	10. $8.21 − 1.85	11. $7.08 − 3.42	12. $9.36 − 5.99

Answer Box

A	B	C	D	E	F
$4.68	$6.46	$2.56	$2.51	$1.56	$5.78
G	**H**	**I**	**J**	**K**	**L**
$3.51	$3.37	$1.64	$6.36	$3.66	$2.77

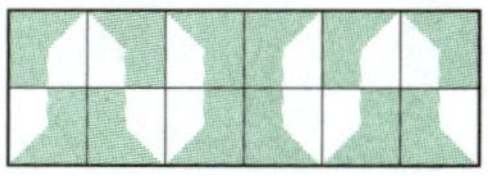

Objective: Find the difference of two amounts of money.

Problem Solving: Choosing the Operation

Children's Art Fair		
Activity	**Saturday**	**Sunday**
Finger painting	325	213
Pottery	225	220
Jewelry making	298	176
Weaving	310	198
Wood carving	300	157

Choose the equation that can be used to solve the problem.

1. How many children made pottery altogether?
   225 + 220 = ■ or 225 − 220 = ■

2. On Sunday, how many more children made jewelry than carved wood?
   176 − 157 = ■ or 176 + 157 = ■

3. Nathan had a $5.00 bill. He paid $2.50 to enter the art fair. How much change did Nathan get?
   $5.00 + $2.50 = ■ or $5.00 − $2.50 = ■

4. How many children took part in weaving at the fair?
   310 − 198 = ■ or 310 + 198 = ■

5. There were enough supplies for 300 children to make a wood carving each day. How many carvings could be made from the supplies left over on Sunday?
   300 + 157 = ■ or 300 − 157 = ■

6. How many more children made pottery on Saturday than on Sunday?
   225 − 220 = ■ or 225 + 220 = ■

7 On Saturday, 198 boys took part in weaving. How many girls took part on Saturday?
310 + 198 = ■ or 310 – 198 = ■

8 On Sunday, 563 children came to the art fair. This was 213 less than the number that came on Saturday. How many children came on Saturday?
563 – 213 = ■ or 563 + 213 = ■

9 On Sunday, how many children made jewelry or carved wood?
176 + 157 = ■ or 176 – 157 = ■

10 Yoko paid $2.50 to get into the art fair. She spent $5.00 for a T-shirt. How much did Yoko spend?
$5.00 – $2.50 = ■ or $5.00 + $2.50 = ■

11 On Sunday, 563 children came to the art fair. How many did not finger paint?
563 + 213 = ■ or 563 – 213 = ■

12 Each child needed a block of wood for wood carving. How many blocks of wood were used?
300 – 157 = ■ or 300 + 157 = ■

Answer Box

A	B	C	D	E	F
300 – 157 = ■	563 – 213 = ■	300 + 157 = ■	563 + 213 = ■	310 – 198 = ■	$5.00 + $2.50 = ■
G	**H**	**I**	**J**	**K**	**L**
$5.00 – $2.50 = ■	176 + 157 = ■	176 – 157 = ■	225 – 220 = ■	310 + 198 = ■	225 + 220 = ■

Objective: Find the addition or subtraction equation that can be used to solve a problem.

Problem Solving: Solving Multi-Step Problems

Solve the problem.

1. Sarah caught 10 fish in the morning. She threw 6 back because they were too small. She caught 3 more in the afternoon. How many fish did Sarah keep?
2. There are 15 rowboats at the lake. There are 10 sailboats. On Saturday, 8 people rent boats. How many boats are not rented?
3. On a nature walk, Ella saw 15 different animals. All except 6 were birds. 8 of the birds were not eagles. How many eagles did Ella see?
4. Ian has $9.00. He spends $2.95 for a map and $4.50 for lunch. How much money does he have left?
5. 63 people came to the lake for a picnic. 29 people left early. 8 more people came. How many people are at the picnic now?
6. 35 adults and 16 children signed up for a nature walk. 4 people did not show up. How many people went on the walk?

7. 20 people are in line to climb the lookout tower. 7 more people get in line. Then 6 people leave. How many are still in line?

8. It costs Mr. Sanders $6.50 to get into the park. He has a coupon for $1.20 off. It costs his son $4.25. How much did Mr. Sanders and his son have to pay?

9. Percy lives 214 miles from the lake. He left home and drove 52 miles before he stopped for gas. Then he drove 68 miles and stopped for a snack. How many miles is Percy from the lake?

10. There are 490 people staying at the lake. 60 people are at the lodge. 96 people are in cabins. The rest are in tents. How many people are in tents?

11. The Olsons spent 40 minutes setting up their campsite and 20 minutes gathering firewood. They swam for 45 minutes. How many more minutes did they spend setting up and gathering wood than swimming?

12. Rachel bought a raft to take to the lake. It cost $21.50. Tax was $1.05. She gave the clerk $30.00. How much change did Rachel get?

Answer Box

A	B	C	D	E	F
$7.45	$9.55	15	42	21	334
G	**H**	**I**	**J**	**K**	**L**
17	47	1	94	7	$1.55

Objective: Solve a multi-step problem involving addition or subtraction.

Do It with Decimals!

Example

Find the sum.

$$\begin{array}{r} 1.6 \\ \underline{+\,0.7} \end{array}$$

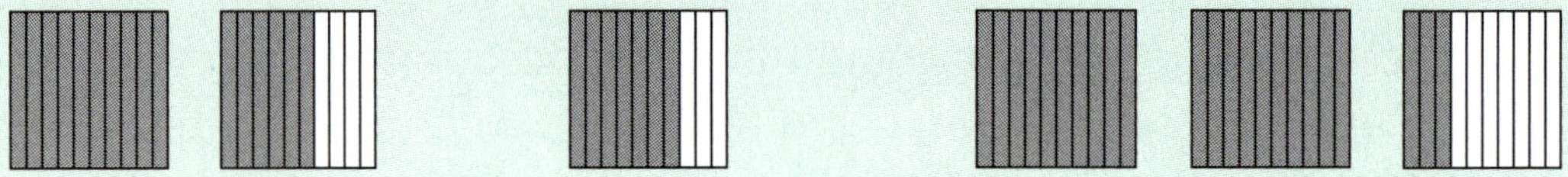

So, 1.6 + 0.7 = **2.3**.

Find the sum or difference.

1. $\begin{array}{r} 0.3 \\ \underline{+\,0.4} \end{array}$

2. $\begin{array}{r} 0.9 \\ \underline{-\,0.3} \end{array}$

3. $\begin{array}{r} 2.2 \\ \underline{+\,3.5} \end{array}$

4. $\begin{array}{r} 0.5 \\ \underline{+\,0.7} \end{array}$

5. $\begin{array}{r} 2.6 \\ \underline{+\,4.8} \end{array}$

6. $\begin{array}{r} 8.9 \\ \underline{-\,1.6} \end{array}$

7. $\begin{array}{r} 9.5 \\ \underline{-\,3.9} \end{array}$

8. $\begin{array}{r} 5.9 \\ \underline{-\,2.1} \end{array}$

9. $\begin{array}{r} 3.5 \\ \underline{+\,3.5} \end{array}$

10. $\begin{array}{r} 3.6 \\ \underline{+\,2.9} \end{array}$

11. $\begin{array}{r} 7.6 \\ \underline{-\,4.8} \end{array}$

12. $\begin{array}{r} 6.2 \\ \underline{-\,1.5} \end{array}$

Answer Box

A	B	C	D	E	F
7.0	5.7	0.6	7.3	0.7	1.2
G	**H**	**I**	**J**	**K**	**L**
7.4	2.8	4.7	3.8	5.6	6.5

 Objective: Find the sum or difference of two decimals in tenths.

Have Some More Hundredths!

Example

Find the sum.

$$\begin{array}{r} {}^{1} \\ 3.45 \\ +\underline{2.17} \\ \mathbf{5.62} \end{array}$$

Regroup with decimals the same way you regroup with whole numbers!

Find the sum.

1. $7.15 + 3.24$
2. $5.72 + 2.29$
3. $2.18 + 3.71$
4. $8.26 + 7.56$
5. $7.24 + 4.87$
6. $4.73 + 3.64$
7. $6.32 + 1.59$
8. $4.77 + 6.34$
9. $4.59 + 2.30$
10. $8.91 + 6.81$
11. $3.58 + 3.79$
12. $2.87 + 7.62$

Answer Box

A	B	C	D	E	F
6.89	5.89	10.49	11.11	10.39	15.82
G	**H**	**I**	**J**	**K**	**L**
12.11	7.37	8.01	8.37	7.91	15.72

Objective: Find the sum of two decimals in hundredths.

Have Fewer Hundredths!

Example

Find the difference.

```
  5 11
  6̸.1̸8
 −4.27
 -----
  1.91
```

Find the difference.

1. $9.49 - 3.03$
2. $6.93 - 4.01$
3. $4.24 - 2.18$
4. $7.17 - 5.25$
5. $8.56 - 4.58$
6. $6.35 - 1.39$
7. $3.52 - 2.37$
8. $8.05 - 2.09$
9. $9.54 - 5.46$
10. $5.20 - 3.95$
11. $7.02 - 4.86$
12. $9.13 - 3.07$

Answer Box

A	B	C	D	E	F
1.92	1.25	6.06	5.96	3.98	4.08
G	**H**	**I**	**J**	**K**	**L**
1.15	6.46	2.92	4.96	2.06	2.16

 Objective: Find the difference of two decimals in hundredths.